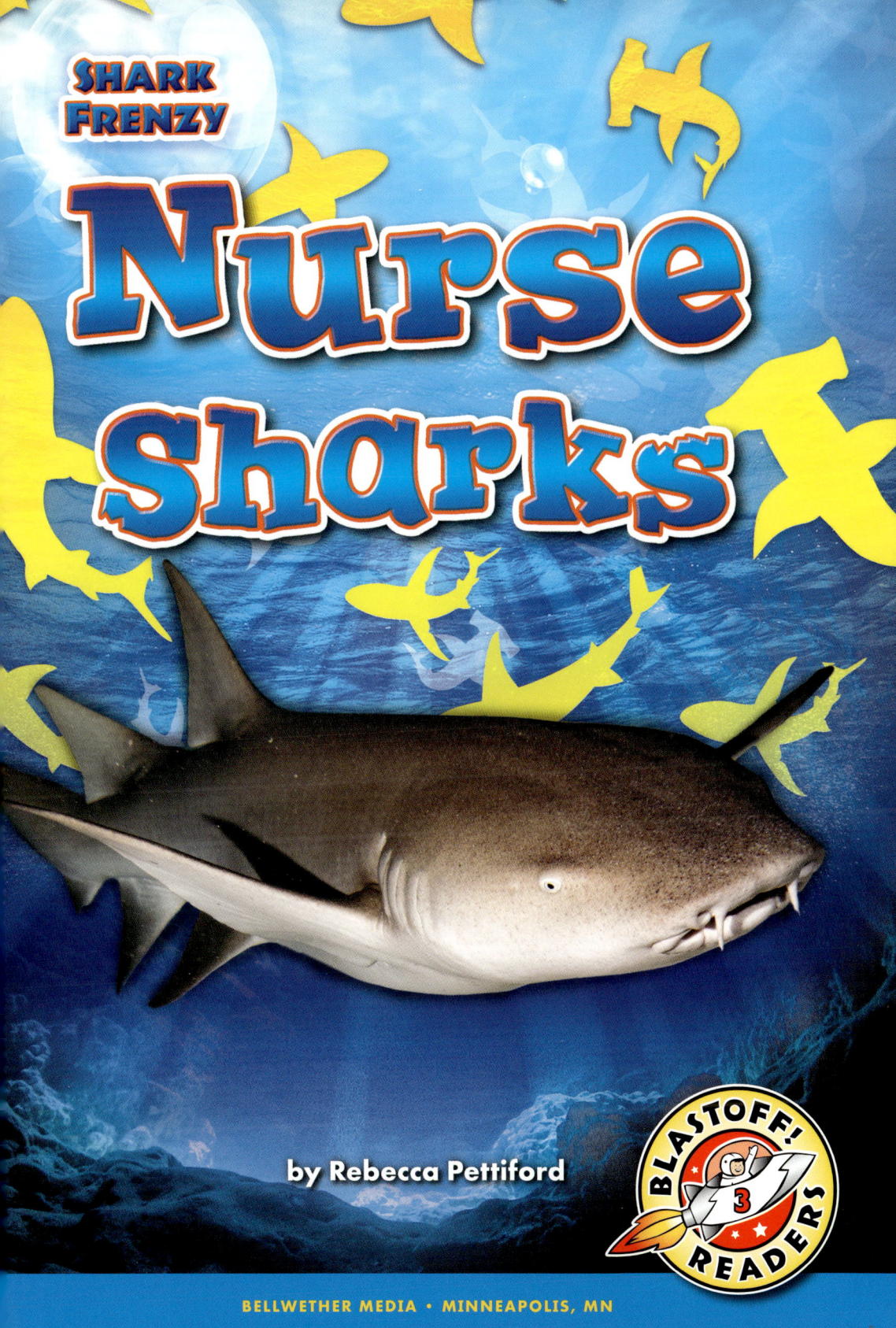

Blastoff! Readers are carefully developed by literacy experts to build reading stamina and move students toward fluency by combining standards-based content with developmentally appropriate text.

 **Level 1** provides the most support through repetition of high-frequency words, light text, predictable sentence patterns, and strong visual support.

 **Level 2** offers early readers a bit more challenge through varied sentences, increased text load, and text-supportive special features.

 **Level 3** advances early-fluent readers toward fluency through increased text load, less reliance on photos, advancing concepts, longer sentences, and more complex special features.

★ **Blastoff! Universe**

Reading Level

 Grade K   Grades 1–3   Grade 4

This edition first published in 2021 by Bellwether Media, Inc.

No part of this publication may be reproduced in whole or in part without written permission of the publisher. For information regarding permission, write to Bellwether Media, Inc., Attention: Permissions Department, 6012 Blue Circle Drive, Minnetonka, MN 55343.

Library of Congress Cataloging-in-Publication Data

Names: Pettiford, Rebecca, author.
Title: Nurse sharks / by Rebecca Pettiford.
Description: Minneapolis, MN : Bellwether Media, Inc., 2021. | Series: Shark frenzy | Includes bibliographical references and index. | Audience: Ages 5-8 | Audience: Grades 2-3 | Summary: "Simple text and full-color photography introduce beginning readers to nurse sharks. Developed by literacy experts for students in kindergarten through third grade"–Provided by publisher.
Identifiers: LCCN 2020036811 (print) | LCCN 2020036812 (ebook) | ISBN 9781644874400 (library binding) | ISBN 9781648341175 (ebook)
Subjects: LCSH: Nurse shark–Juvenile literature.
Classification: LCC QL638.95.G55 P48 2021 (print) | LCC QL638.95.G55 (ebook) | DDC 597.3/3-dc23
LC record available at https://lccn.loc.gov/2020036811
LC ebook record available at https://lccn.loc.gov/2020036812

Text copyright © 2021 by Bellwether Media, Inc. BLASTOFF! READERS and associated logos are trademarks and/or registered trademarks of Bellwether Media, Inc.

Editor: Rebecca Sabelko   Designer: Josh Brink

Printed in the United States of America, North Mankato, MN.

Table of Contents

What Are Nurse Sharks?	4
Bottom-dwellers	8
Pile of Sharks	14
Deep Dive on the Nurse Shark	20
Glossary	22
To Learn More	23
Index	24

What Are Nurse Sharks?

Nurse sharks are found in the warm waters of the Atlantic and Pacific Oceans.

They swim near **coral reefs** and rocky coasts. They also live near **mangrove** islands.

Nurse Shark Range

range =

In some areas, people hunt nurse sharks for their fins and meat.

scientists studying a nurse shark

Scientists believe the nurse shark population is falling. They want to learn more to help these sharks.

Bottom-dwellers

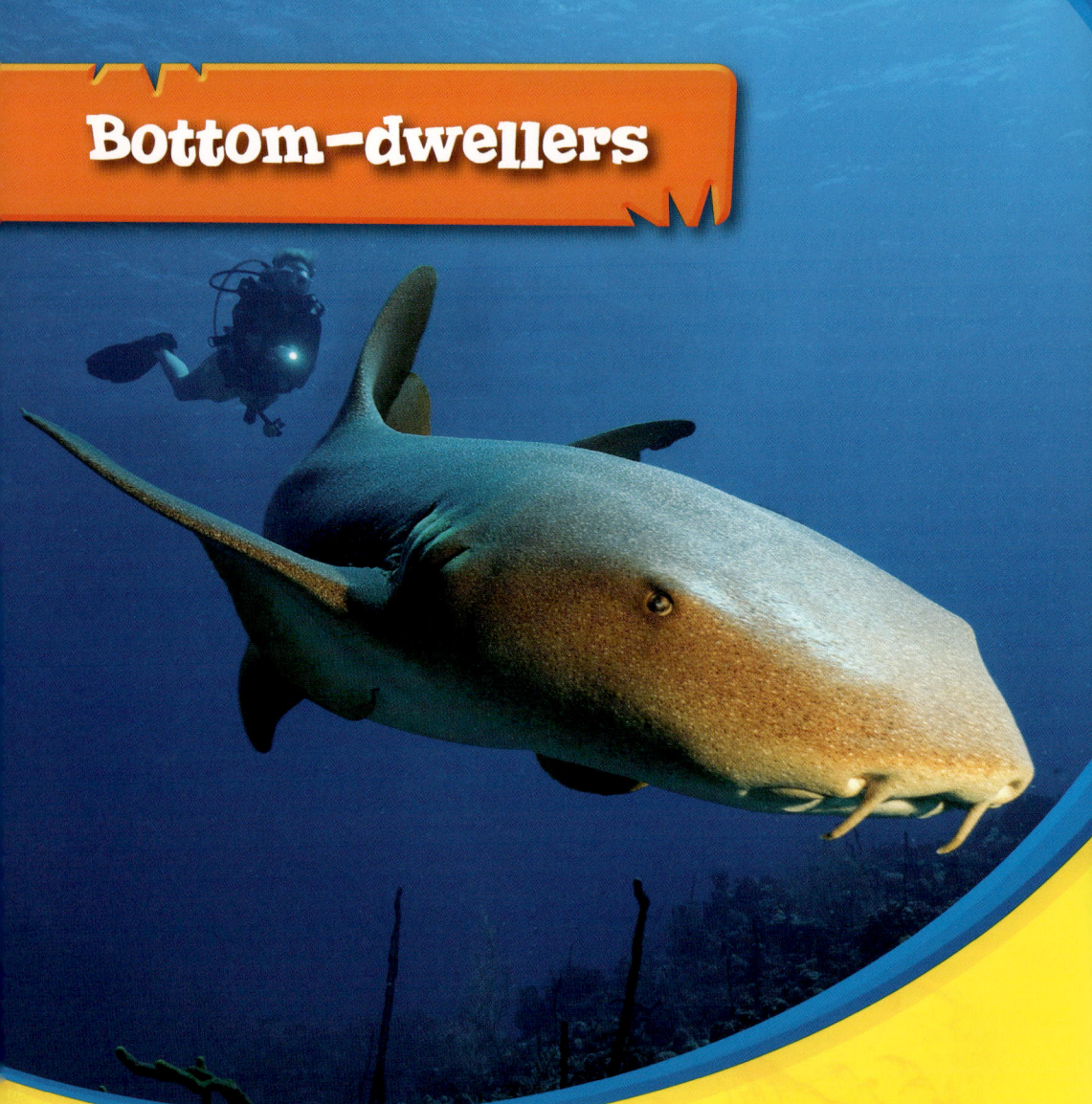

Nurse sharks have flat bodies. They can be yellow or brown.

They can reach up to 14 feet (4.3 meters) long. But most are around 8 feet (2.4 meters) long.

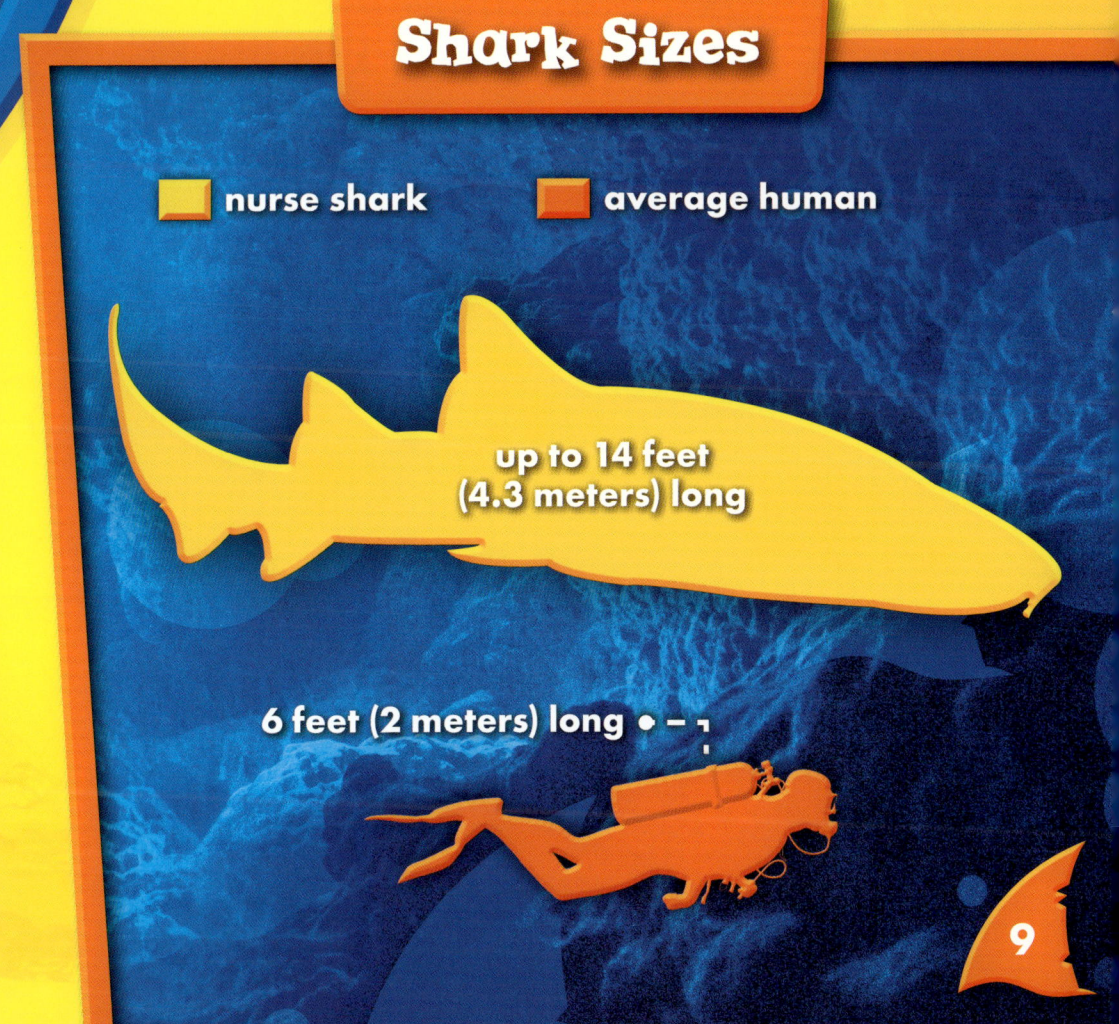

Shark Sizes

■ nurse shark   ■ average human

up to 14 feet (4.3 meters) long

6 feet (2 meters) long

Nurse sharks are **bottom-dwellers**. They use their **pectoral fins** to hold their bodies off the ocean floor.

Small **prey** hide under their bodies. Nurse sharks attack these fish!

pectoral fin

Nurse sharks have **barbels** that stick out next to their mouths. They brush their barbels across the sandy floor to find food.

Identify a Nurse Shark

round snout

flat body

barbels

Their small mouths are located under their round **snouts**. This allows the sharks to easily suck up their prey.

Pile of Sharks

During the day, nurse sharks rest in groups. Groups can include up to 40 sharks. Sometimes they pile on top of each other!

At night, nurse sharks hunt alone. They search for small fish, shellfish, and stingrays.

Nurse Shark Diet

small fish

shellfish

stingrays

Nurse sharks hunt by pushing their snouts against the sand and small holes in rocks. Then they suck up food.

Nurse sharks have powerful mouths. They can suck a whole **conch** from its shell!

Nurse sharks do not have many natural **predators**. But lemon sharks and tiger sharks may attack them.

Nurse sharks are one of the ocean's **unique** swimmers!

Deep Dive on the Nurse Shark

 **LIFE SPAN:**
up to **25 years**

 **LENGTH:**
up to **14 feet (4.3 meters) long**

 **WEIGHT:**
up to **330 pounds (150 kilograms)**

 **DEPTH RANGE:**
0 to 427 feet (0 to 130 meters)

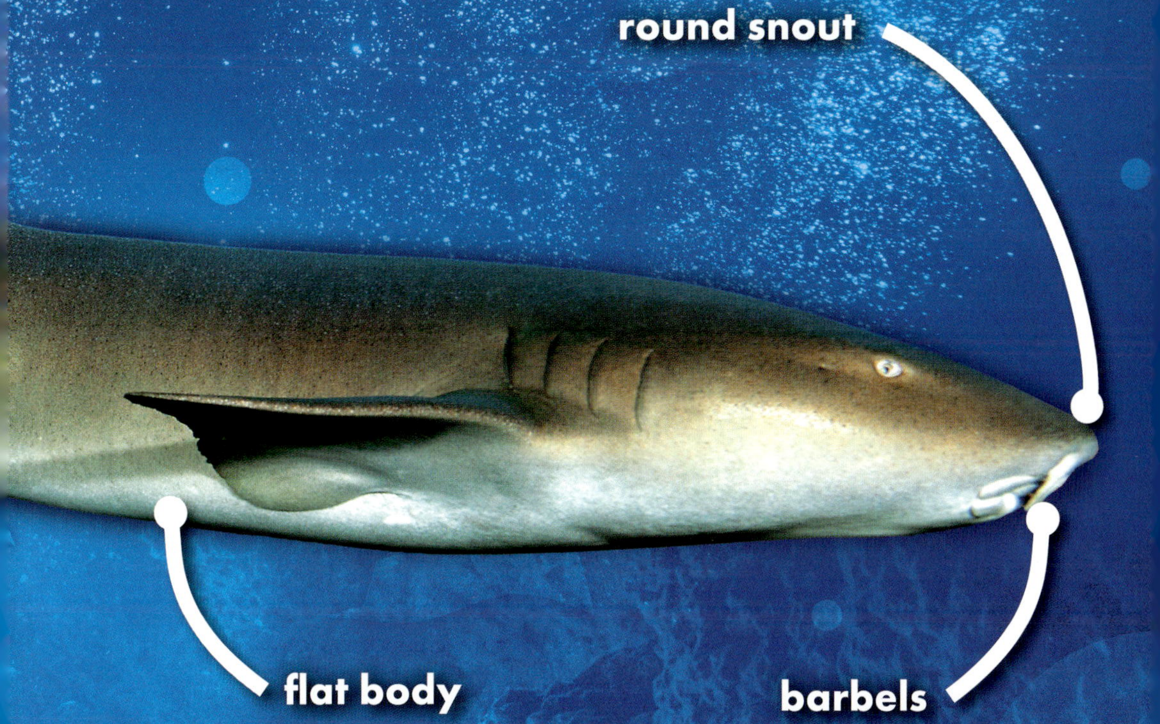
round snout
flat body
barbels

Least Concern	Near Threatened	Vulnerable	Endangered	Critically Endangered	Extinct in the Wild	Extinct

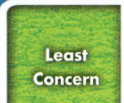

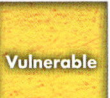

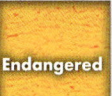

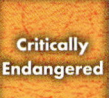

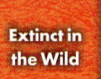

conservation status: unknown

Glossary

barbels—whisker-like body parts around the mouths of nurse sharks that are used to find food

bottom-dwellers—animals that spend a lot of time on or near the bottom of the ocean

conch—a type of shellfish that lives in a large shell

coral reefs—structures made of coral that usually grow in shallow seawater

mangrove—related to groups of trees that grow in shallow saltwater or swamps

pectoral fins—fins on the sides of sharks that control movement

predators—animals that hunt other animals for food

prey—animals that are hunted by other animals for food

snouts—the noses of some animals

unique—one of a kind

To Learn More

AT THE LIBRARY

Hansen, Grace. *Nurse Sharks.* Minneapolis, Minn.: Abdo Kids, 2016.

Markle, Sandra. *What If You Could Sniff Like a Shark? Explore the Superpowers of Ocean Animals.* New York, N.Y.: Scholastic Press, 2020.

Murray, Julie. *Nurse Sharks.* Minneapolis, Minn.: Abdo Zoom, 2020.

ON THE WEB

FACTSURFER

Factsurfer.com gives you a safe, fun way to find more information.

1. Go to www.factsurfer.com.
2. Enter "nurse sharks" into the search box and click 🔍.
3. Select your book cover to see a list of related content.

Index

Atlantic Ocean, 4
barbels, 12, 13
bodies, 8, 10, 13
bottom-dwellers, 10
color, 8
coral reefs, 5
deep dive, 20-21
food, 10, 12, 13, 15, 16
hunt, 6, 15, 16
mangrove islands, 5
mouths, 12, 13, 16
ocean floor, 10, 12
Pacific Ocean, 4
pectoral fins, 10, 11
pile, 14
population, 7
predators, 18
prey, 10, 13, 15, 16
range, 4, 5

rest, 14
scientists, 7
size, 9
snouts, 13, 16
swim, 5, 19
waters, 4

The images in this book are reproduced through the courtesy of: Andrea Izzotti, front cover (hero), p. 13; Daniel Lamborn, p. 3; Katie Thorpe, p. 4; Nature Picture Library/ Alamy, p. 6; Danita Delimont/ Alamy, p. 7; jimcatlinphotography.com, pp. 8, 18; WaterFrame_fba/ Alamy, p. 10; WaterFrame_mus/ Alamy, pp. 10-11; Michael Bogner, p. 12; Vlad61, p. 13 (ocean floor); Maui Tropical Images, p. 13 (call out); Water Frame/ agefotostock, p. 14; Danny Ye, p. 15 (top left); DPFishCo, p. 15 (top right); Drew McArthur, p. 15 (bottom); RVLIKEMIDGLEY, pp. 16, 16-17; Matt9122, p. 19; Beat J Korner, pp. 20-21; Tatiana Belova/ Dreamstime, p. 23.